Do Just One thing for the Sea...

The Irish Sea for Society Compendium

A handy miscellany and all round sea-loving production

Authors – Niamh Dornan, Tanja Calis, John Joyce, Trevor Purtill
Illustrators – Lluís C. Catchot, John Joyce

First edition 2015
ISBN 978-0-9926602-1-5
Published by AquaTT
©AquaTT 2015

The Sea for Society Mobilisation and Mutual Learning (MML) project is supported by the European Commission under the Science in Society Theme of the 7th Framework Programme for Research and Technological Development under grant agreement no. 289066. This publication only reflects the views of the author(s), and the European Union cannot be held responsible for any use which may be made of the information contained therein.

www.aquatt.ie

Contents

About this Book	1-2
Do Just One Thing	3-6
The Sea As: A Place to Live	7-8
The Sea As: A Source of Food	9-10
The Sea As: A Means of Transport	11-12
The Sea As: A Source of Energy	13-15
The Sea As: A Support for Human Health	16-17
The Sea As: A Place for Leisure & Tourism	18
Irish Sea Heroes	19-25
Books About the Sea	26-28
Summertime Shanty	29-42
Sea Heroes: Jacques-Yves Cousteau	43
Great Shipwrecks in History: Lusitania	41
Great Shipwrecks in History: Titanic	45
Sea Heroes: Robert Ballard	45-46
Great Shipwrecks in History: Spanish Armada	47
10 Songs about the Sea	48-50
10 Films about the Sea	54-55
Sea Creatures	53-56
Sea Heroes: Eugenie Clark	57-58
Sea for Society Quiz	59-60
Colouring Sheets	61-64
Quiz Answers	65
Do Just One Thing Checklist	66

About this Book

Sometimes we don't realise how crucial our seas are, and sometimes we act irresponsibly towards them. As a society, our consumption behaviours are threatening the health of our seas, leading to their gradual deterioration through pollution, ocean acidification, waste dumping and overfishing.

Now it is time for us to stop taking our seas for granted. Try to imagine where we would be without our seas. Half of the oxygen we breathe is produced by marine phytoplankton. Our seas moderate our weather and provide the main protein source for over a billion people. Yet still we mistreat the seas.

This book has been produced as part of an extensive European Commission-funded project called Sea for Society. We are trying to address the threats posed to the seas by human behaviour, but we are also exploring and promoting the many opportunities ours seas can provide, including renewable energy, sustainable food supply, and climate regulation.

The project is now in its mobilisation phase. We are inviting you to come on a journey with us, a journey that is being replicated in many countries across Europe, a journey in which we will explore what the seas mean to our lives and what we can do to ensure we live in sustainable harmony with the seas.

A key element of the project so far has been the public consultation on six themes:
- The Sea as a Place to Live
- The Sea as a Source of Food
- The Sea as a Means of Transport
- The Sea as a Source of Energy
- The Sea as a Place for Leisure and Tourism
- The Sea as a Support for Human Health

Each of these themes is further explored in this book. We would like you to consider these themes and reflect on how they are relevant to your life.

We are also asking you to **Do Just One Thing for the Sea** by completing one or more of the actions listed on pages 3-5. These actions represent simple and achievable steps that, if collectively taken, can have a huge beneficial impact on our seas.

Elsewhere in this book we have features on Irish and international sea heroes, interesting sea-related histories, some fun lists, a comic, and plenty of other interesting features about our seas – just dive in and start discovering!

www.seaforsociety.eu

www.seaforsociety-irelandblogspot.ie

 @SeaforSocietyIE

Do Just One Thing...
Ten Simple Actions You Can Take To Help Protect Our Seas

1. Dispose of plastic products responsibly – do not throw them in the sea or leave them on the beach

Why? Plastic products do not dissolve in water but remain in large or small pieces literally forever. Large pieces, such as bottle separators and fishing lines trap marine animals. Smaller pieces may be eaten by marine mammals, fish and birds - choking them to death. And all plastic products can be ground down by the ocean into microscopic particles that may clog the digestive systems of marine animals, gradually starving them to death.

2. Volunteer to take part in a local beach clean

Why? Rubbish left on beach is unsightly and can be dangerous both to wildlife and other beach users.

3. Only eat fish that has been caught from 'Sustainable Fisheries' (look for the declaration on the packet or ask your local fishmonger and, when dining out, always check with restaurants that the fish they serve is from a sustainable fishery)

Why? Unsustainable fishing, which reduces fish populations to the point where there are not enough adult fish left to repopulate the stock, destroys the fish resource for generations to come.

4. Always read the labels on pet food containers to make sure that it contains only sustainably caught marine proteins

Why? Protein from unsustainable fisheries leads to overfishing, which can destroy whole marine ecosystems.

5. Leave your car at home

Why? Burning petrol and diesel releases carbon dioxide into the atmosphere which, when absorbed into the sea, increases its acidity. Acid water dissolves the chalky shells that protect a wide range of animal species – including lobsters, crabs, prawns, shrimps, oyster, scallops, corals and a host of microscopic animals.

6. Turn down your heating

Why? Burning fossil fuels such as wood, oil, coal and natural gas to create electricity increases the amount of 'greenhouse gases' such as water vapour, carbon dioxide, methane, nitrous oxide and ozone. This has the effect of trapping heat within the atmosphere, which in turn leads to an increase in ocean temperature. As the oceans warm, ocean circulation is effected, polar ice melts and sea levels rise – causing flooding and storms.

7. Have a shower instead of a bath

Why? Showers use less fresh water than a bath. This means that less clean, fresh water has to be generated to provide the experience in the first place and less dirty water has to be disposed off afterwards (usually through sewage systems into the ocean).

8. Don't buy holiday souvenirs that exploit the oceans such as dried seahorses, seashells and corals

Why? Corals and sea shells can take decades to grow and only a few seconds to destroy.

9. Learn more about the sea

Why? The sea covers 70% of our planet, provides half of the oxygen that we breathe and regulates the world's weather. It is – quite literally – the life support system of planet Earth. Learning about the sea and how it works is not only essential information for every citizen of this planet but also a wonderland of fascinating facts and statistics, stories, songs and poetry. This book is a great place to start!

10. Spread the word

Why? Learning something from someone we know and trust is far more likely to call us to action than something we read in a book or see on television. So tell all your friends, your relatives and anyone who will listen about the wonderful resource that is the sea and the wealth of entertaining stories it has to tell.

The Sea As: A Place to Live

Without the ocean, life as we know it would not exist!

The ocean makes Earth a habitable place to live. Almost every living thing needs oxygen to survive. All oxygen gas on Earth originally came from organisms in the ocean that photosynthesise, just like terrestrial plants do on land. The earliest evidence of life is found in ancient sea sediments and the millions of different species of organisms on Earth today descended from common ancestors that evolved in the ocean.

The ocean itself is also a place to live. It is now estimated that around 50-80% of all life on Earth is found in the sea and presently about 40% of the world's population live within 100km of the coast.

Ireland's coastline is 7,000km long, yet no place in the whole country is more that 100km from the coast!

The Life Aquatic:

Some intrepid ocean explorer's have even chosen to live under the sea!

Aquarius Reef Base - The Aquarius Reef Base is the world's only remaining undersea research laboratory, located 18m beneath the ocean's surface in the Florida Keys National Marine Sanctuary. Since its deployment in 1993, 'aquanauts' and their support teams have used Aquarius to carry out more than 120 missions to research coral reefs, ocean acidification, climate change, fisheries and the overall health of the oceans. Aquanauts diving from Aquarius can spend nine hours per day diving to depths of 30m with a reduced risk of decompression sickness ('the bends').

In 2014, French aquatic filmmaker and oceanographic explorer Fabien Cousteau spent 31 days underwater in Aquarius in a tribute to his grandfather Jacques. In doing so, he set the record for longest time underwater for a film crew, surpassing his grandfather's 30 days, and collected a large amount of scientific data.

Learn more at: www.aquarius.fiu.edu

> **"The Sea, once it casts its spell, holds one in its net of wonder forever."**
>
> Jacques Yves Cousteau, Oceanographer

The Sea As: A Source of Food

After cereals, fish and other seafood are some of the most important foods consumed worldwide. They provide about 15% of the world population's protein intake, and at least 50% of total animal protein intake in some small island developing states.

Fishing is one of the oldest activities of humankind. Ancient heaps of discarded mollusk shells, some from prehistoric times, have been found in coastal areas throughout the world, including those of China, Japan, Peru, Brazil, Portugal, and Denmark, indicating that seafood was part of the earliest human diets.

Irish Seafood Industry

The Irish seafood industry contributes about €700 million annually to national income and employs 11,000 people, mainly in coastal counties.

The four main activities in the Irish seafood industry are:
- Fishing – The top fishing ports in Ireland are Killybegs, Castletownbere, Dingle, Dunmore East and Kilmore Quay, but fishing vessels also land into numerous small ports around the coast.
- Aquaculture (fish farming) - Aquaculture activity includes growing finfish, such as salmon and trout, and shellfish farming, including the cultivation of mussels, oysters and scallops.
- Processing - Seafood companies produce high value products from salmon, whitefish, shellfish, herring, and mackerel.
- Marketing - Irish seafood is sold at home (€340 million) and in international markets (Europe, Africa and the Far East) where exports are valued at €375 million.

Bia Bocht

Certain shellfish can literally be picked up on the beach at low tide. Up until the nineteenth century it was mainly the poor who gathered shellfish. Such fare, called 'cnuasach mara' (sea pickings) was known as 'bia bocht' or poor man's food.

The Sea As: A Means of Transport

Evidence of sea-going vessels date back over 9,000 years and the oldest European boat is a simple 3m long dugout dated to 7400 BC, discovered in Pesse, Holland.

In modern times, the sea has become the dominant and most cost-effective means for the transport of materials, transporting an estimated 95% of the world's traded goods. In 2012, approximately 13,000 passengers travelled to and from Ireland by sea.

Currach

A currach is a type of Irish boat with a wooden frame, over which animal skins or hides were once stretched, though now canvas is more usual. It is sometimes anglicised as "Curragh". The construction and design of the currach are unique to the west coasts of Ireland and Scotland, with variations in size and shape by region. The plank-built rowing boat found on the west coast of Connacht is also called a currach or curach adhmaid ("wooden currach"), and is built in a style very similar to its canvas-covered relative. A larger version of this is known simply as a bád iomartha (rowing boat).

The currach has traditionally been both a sea boat and a vessel for inland waters.

The Sea As: A Source of Energy

The ocean is an enormous source of energy. It is estimated that less than 1% of the energy in ocean waves could be capable of supplying 500% of the entire world's energy requirements. Today, ocean energy covers around 0.02% of EU energy needs and it is primarily used for electricity production.

Ireland's geographical position and climate mean that it is one of the best locations in the world for marine renewable energies.

Tide

Tidal energy involves the harnessing and converting of the kinetic or moving energy present in the inflow or outflow of tidal waters into a more usable form of energy i.e. electricity.

There are two main types of tidal harvesting technology:

- **Tidal barrage systems** involve the trapping of water at high tide, often near the mouth of a river estuary or a bay, followed by the controlled release of the water back to the sea through turbines.
- **Tidal stream turbines** harness the power of both the inflows and outflows of tidal energy. The tide flows through the stationary turbines, causing them to turn using the same principal as a wind turbine.

Wave

There are many different designs and methods being used to harness wave energy, including:

- **Attenuators** are long floating structures made up of different segments, which are placed in parallel (same direction) to the direction of the waves. As the device is lifted and dropped at the same time at different points on the structure, this causes spreading out (expansion) and tightening (contraction) where the segments are connected. This flexing motion can then power hydraulic pumps or other converters which may then power a generator.

- **Point absorbers** operate on the basis of using the bobbing motion of the wave at a fixed point to cause motion of components within, or attached to a floating structure, such as a floating buoy, within a fixed cylinder. This stretching and compressing motion can then be used to drive electromechanical or hydraulic energy converters.

- **Termination** wave devices involve the physical capturing or reflection of the power of the wave. The "oscillating water column" is a form of device in which water enters through an opening into a chamber with air trapped above it. The wave action causes the captured water to move up and down a column like a piston, forcing air though an opening connected to a turbine which then generates electricity.

Off-shore wind

We can harness wind energy and convert it into useful electrical energy using wind turbines. Instead of using electricity to make wind, like a fan, wind turbines use wind to make electricity. The wind turns the blades, which spin a shaft, which connects to a generator and makes electricity. Wind turbines are often combined in groups or "wind farms". Wind farms can be located on land (on-shore) or out at sea (off-shore).

Off-shore wind farms are present off the shores of numerous countries including off the east coast of Ireland.

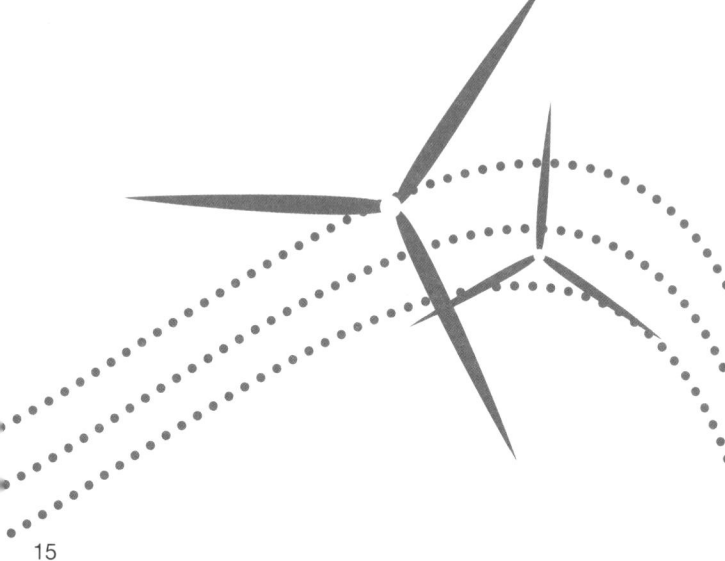

The Sea As: A Support for Human Health

The marine environment can be the source of potential human health benefits through the provision of healthy food, novel pharmaceuticals and related products derived from marine organisms, as well as through a contribution to general well-being from a close association with the coastal environment.

Holidays by the seaside, sunbathing and seaweed baths and have traditionally been considered to be beneficial and are all part of a holistic approach to health. Victorian doctors often prescribed periods of recovery in seaside towns after serious illness.

Marvellous Marine Medicine

By studying marine animals, plants, fungi and bacteria, scientists are discovering new compounds and uncovering clues that are helping cure disease and improve human health.

Products derived from marine species have provided chemotherapy agents, antibiotics, anti-virals, anesthetics, adhesives, marine genetic products and others that being used or developed to treat cancer, leukemia, cystic fibrosis, heart disease, wounds and infections, among others.

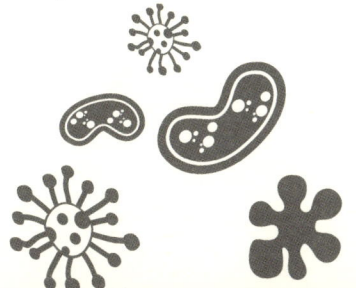

Did you know?

- Compounds used to fight the AIDS virus have been discovered in sea sponges
- Soft corals produce a group of compounds with anti-inflammatory properties, which are used in anti-wrinkle cream
- Biologists trying to unravel the puzzles of night blindness and inherited eye disorders have been studying the retinas of skates (a type of flatfish)
- Corals and mollusks are used to make orthopedic surgical implants.

Jellyfish Help Illuminate Cells

The *Aequorea Victoria* jellyfish has light-emitting organs which contain the green fluorescent protein (GFP) which glows intensely under ultraviolet light. The green light enables scientists to track, amongst other things, how cancer tumours form new blood vessels, how Alzheimer's disease kills brain neurons and how HIV infected cells produce new viruses. Osamu Shimomura, Martin Chalfie, and Roger Tsien were awarded the 2008 Nobel Prize in Chemistry for the discovery and development of GFP.

The Sea As: A Place for Leisure & Tourism

The seaside resort and beach holiday played a central role in the development of tourism as a great international industry across the globe and today tourism is an economic backbone of many coastal regions. According to recent figures, around 2.36 million people are currently employed in the coastal tourism sector, representing 1% total EU employment. Additionally, more than half of all EU hotel beds are located in regions with a sea border.

Water sports are very popular all around the Irish coastline. From adrenaline-fuelled activities such as kite surfing and wakeboarding, to leisurely paddle boarding, there are options to suit everyone. Irish waters are teeming with life and fascinating shipwrecks, which also makes Ireland is a great location for dive enthusiasts. Believe it or not Ireland is also a top global destination for surfing thanks to our coastline's range of beach, reef and break points and world famous giant waves! Kayaking and sailing also offer unique ways to explore the Irish coastline.

Irish Sea Heroes: Brendan the Navigator
(c484 - c597)

The legend that St. Brendan the Navigator beat the Vikings to the discovery of America was reinforced in 1983 by the discovery of writings in Irish Ogham script that described the Christian nativity, carved into rock in West Virginia.

While scholars have disputed this discovery, Viking sagas tell of Native Americans who had already encountered explorers dressed in white from a land "across from their own". Another tribe was reported to have spoken in a language that sounded like Irish, which the Vikings knew and recognised. To establish if such a voyage could have been possible, the explorer Tim Severin recreated St. Brendan's boat and, between May 1976 and June 1977 sailed it successfully across the Atlantic from Ireland to Newfoundland, stopping at the Hebrides and Iceland on the way.

> **A boat sailed out of Brandon
> in the year of 501**
>
> **He ploughed a lonely furrow to the north, south, east and west**
>
> **Of all the navigators, St. Brendan
> was the best.**
>
> From *St Brendan's Voyage* by Christy Moore

Irish Sea Heroes: Grace O'Malley
(c1530 – c1603)

Grace O'Malley, the famous 16th century 'Sea Queen of Ireland', got her nickname 'Gráinne Mhaol' – 'bald Gráinne' – following an argument with her father when she shaved off most of her long hair after he said it would get caught in the ship's ropes if he allowed her to come on a trading expedition to Spain.

Gráinne became a legendary figure in the west of Ireland – as a seafarer and even as a 'pirate' – demanding 'tax' from seafarers who traded around Mayo (although such taxes were also levied by Galway). She met with Queen Elizabeth I in 1593 to successfully petition for the release of her sons and half-brother who had been captured by the English governor of Connaught. Both women spoke in Latin, since Elizabeth spoke no Irish and Gráinne spoke no English.

> 'Twas a proud and stately castle
> In the years of long ago
> When the dauntless Grace O'Malley
> Ruled a queen in fair Mayo.
> And from Bernham's lofty summit
> To the waves of Galway Bay
> And from Castlebar to Ballintra
> Her unconquered flag held sway.
>
> From *Granuaile* – circa 1798 – available in Hardiman's Irish Minstrelsy vol. II

Irish Sea Heroes: Francis Beaufort
(1774-1857)

Irish hydrographer Francis Beaufort was far from being a backroom scientist. Prior to his development of the 'Beaufort Scale' of wind measurement, he was severely wounded in the left arm and chest by sword and blunderbuss during his time in the Royal Navy. Later on, during a mapping expedition in the Mediterranean he was wounded a second time by a musket ball fired by a band of Turks.

In 1829, Beaufort was appointed to run the Hydrographic Office in Greenwich, which he transformed into the greatest collection of hydrographic data in the world. In 1831, Beaufort wrote to his friend Commander Fitzroy of the survey ship HMS *Beagle*, recommending 'A Mr. Darwin' whom, without Beaufort's recommendation, may never have changed the foundation of biology.

Irish Sea Heroes: John Phillip Holland (1840 – 1914)

Holland was a former teacher and Christian Brother from Liscannor, Co. Clare who emigrated to America in 1873. In May 1879, after having his submarine designs rejected by the US Navy, he began work on *The Fenian Ram* - a primitive thirty-foot long submarine. This was paid for from the 'Skirmishing Fund' of the Fenian Brotherhood to attack the British Navy. But arguments over the cost and disagreement in the ranks of the Brotherhood meant the 'Ram' was never used in anger. Holland went on to design bigger and better submarines including the Royal Navy's first underwater vessel, HMS *Holland*. The company he founded – The Electric Boat Company – grew to become the major submarine contractor General Dynamics - one of the leading suppliers of nuclear submarines to the US Navy.

Irish Sea Heroes: Easkey Britton

Easkey Britton is an internationally renowned, pioneering big-wave surfer from Ireland. She learned to surf when she was just four years old and is five times Irish National Surfing Champion, a past UK Tour Champion and was recently nominated for the Billabong XXL awards.

Her passion for the marine environment led her to co-found the Wellcoast network in 2010 and to complete a PhD in Environment and Society specialising in human wellbeing and coastal resilience. In 2013, Easkey co-founded Waves of Freedom, an initiative founded on the belief in surfing as a powerful medium for creating positive social impact and empowerment, especially for more vulnerable groups like youth and women.

Books About the Sea

Fiction

- *The Odyssey* (c700BC) by Homer (Poem) Odysseus fights off monsters and gods while his wife fights off suitors and pretenders to the throne.
- *Twenty Thousand Leagues Under the Sea* (1869) by Jules Verne The ultimate underwater adventure, groundbreaking Disney movie too!
- *Moby Dick* (1851) by Herman Melville Before 'Jaws' was the Great White Whale.
- *The Old Man and the Sea* (1952) by Ernest Hemmingway Plucky pensioner on his lone quest to hook a giant fish.
- The *'Hornblower'* books (1937-1967) by C.S.Forester Follows a young man from midshipman to Admiral in the British Napoleonic wars.
- The *'Aubrey/Maturin'* books (1969-2004) by Patrick O'Brien Wonderful 'reboot' of Hornblower-type stories.
- *The Sea* (2005) by John Banville Booker Prize-winning fictional memoir.
- *The Open Boat* (1897) by Stephen Crane (Short Story) Based on the author's experience of being shipwrecked at sea.
- *The Rime of the Ancient Mariner* (1798) by Samuel Taylor Coleridge (Classic Poem) Legendary poem which makes you thirsty just reading it.
- *The Mermaid's Purse* (1993) by Ted Hughes (Collection of Poems for children) 28 poems about the sea and sea creatures.

Non-Fiction

- ***The Sea Around Us* (1951) by Rachel Carson**
  Prize-winning best seller.
- ***The Silent World* (1956) by Jacques Yves Cousteau**
  Book (and later the film) that spawned the Cousteau legend.
- ***The Depths of the Sea* (1873) by R. Wyville Thompson** Revolutionised scientific thinking about the deep sea.
- ***Seven-Tenths* (1992) by James Hamilton-Patterson**
  Thought-provoking essays on humankind's fascination with the sea.
- ***Two Years Before the Mast* (1840) by Henry Richard Dana** Classic tale of a two-year voyage in 1983.
- ***The Sea and Civilisation: A Maritime History of the World* (2013) by Lincoln Paine** Detailed exploration of our relationship with the sea.
- ***Kon-Tiki – Across the Pacific by Raft* (1990) by Thor Heyerdahl** Landmark adventure of the 20th century.
- ***In Harm's Way: The Sinking of the U.S.S. Indianapolis and the Extraordinary Story of its Survivors* (2001) by Doug Stanton** Four days' immersed in shark infested waters - relived by the crusty fisherman Quint in 'Jaws'.
- ***A Night to Remember* (1955) by Water Lord**
  The original book on the sinking of the *Titanic.*
- ***Wish for a Fish: All About Sea Creatures (Cat in the Hat's Learning Library)* (1999) by Bonnie Worth**
  Or should that the 'The Catfish in the Hat'?

1.

A day like every other. Jim, the intrepid and unwitting hero of our story, is enjoying a day beside the sea.

2.

BUT LO!

What is this we spy? ...A single plastic bag cartwheeling away and cantering toward the sea...

3.

Jim is too preoccupied with flying his kite to stop the bag going into the water.

MARINE LITTER IS A HUGE CONCERN AND POSES A REAL THREAT TO ALL TYPES OF SEA CREATURES...
...
..PLASTIC RINGS FROM PACKS OF SOFT DRINKS CAN STRANGLE SEA BIRDS...
DISCARDED PLASTIC BAGS ARE ALSO A SOURCE OF REAL DANGER FOR SEA ANIMAL...

4.

THE NEXT DAY, JIM, A KEEN SCHOLAR AS EVER THERE WAS, IS SLEEPING HIS WAY THROUGH CLASS. HIS TEACHER IS DESCRIBING THE EFFECTS OF MARINE LITTER. TOPICAL AND CONVENIENT, YOU SAY? WELL THAT'S HOW THIS STORY IS GOING TO ROLL.

5.

MISS STEPHENS: MARINE LITTER IS A HUGE CONCERN AND POSES A REAL THREAT TO ALL TYPES OF SEA CREATURES AND DISCARDED PLASTIC BAGS ARE ALSO A SOURCE OF DANGER FOR SEA ANIMALS.

6.

JIM:
ZZZZZZ
ZZZZZZzz
zzzzzzzzzzzz
zzzzzzzzzzz
zzzzzzzzzzz
zzzzzzzzzzz
zzzzzzzzzzz

7.

LATER THAT EVENING, JIM IS HAVING SUPPER WITH HIS PARENTS.

HIS MOTHER AND FATHER NOTICE THAT HE IS QUIETER THAN USUAL, AS IF HE IS PREOCCUPIED BY SOMETHING.

8.

THEY ARE CORRECT IN THINKING THAT HIS THOUGHTS HAVE STRAYED FROM HIS USUAL CONCERNS, NAMELY SPORT AND FEEDING HIMSELF. SOME OF MISS STEPHENS' WORDS ARE TRUNDLING AROUND HIS BRAIN.

ZZZ... ZZ
WAKEY, WAKEY! WE'RE GONNA GO ON A VISIONQUEST TOGEEETHER!

9. JIM, EXHAUSTED FROM HIS MILD BOUT OF RETROSPECTION, GOES TO BED EARLY. HE IS WOKEN BY A STRANGE AND UNSETTLING VOICE.

10. **VOICE:** HEY, JIM. WAKEY, WAKEY! OL' SCRIMSHAW JONES WANTS A WORD WITH YA!

11. JIM WAKES UP IN ALARM AND SEES A TRULY VILE AND EVIL LOOKING PIRATE CAPTAIN STANDING AT THE FOOT OF HIS BED.

12. **JIM:** WHAT'S GOING ON? !! I'M GOING TO CALL MY MOM!! WHAT ARE YOU DOING HERE????

SCRIMSHAW JONES: WELL, JIM, LET'S JUST SAY THAT I REPRESENT YOUR CONSCIENCE.

JIM: SOME CONSCIENCE! A THIRD RATE PIRATE, MORE LIKE!

13. **SCRIMSHAW JONES:** YOUR CONSCIENCE CAN TAKE MANY FORMS, KID. SINCE THIS IS ABOUT THE SEA, YOUR CONSCIENCE HAS CREATED A COMPOSITE OF ALL THAT YOU ASSOCIATE WITH THE SEA. IN THIS CASE, A PIRATE. TA DA!

14. **JIM:** WHAT?!!?

SCRIMSHAW JONES: HEY, NO ONE EVER CLAIMED THAT IMAGINATION WAS YOUR STRONG SUIT.

15.

JIM: IF YOU ARE MY CONSCIENCE AND NOT SOME MURDEROUS PSYCHOPATH THAT RUNS A FANCY DRESS SHOP IN HIS SPARE TIME, WHAT EXACTLY ARE YOU DOING HERE?

16.

SCRIMSHAW JONES (SMILING): GLAD YOU ASKED! WE'RE GONNA GO ON A VISION QUEST TOGETHER...

JIM: OH BOY...

17. **SCRIMSHAW JONES:** ... AND SEE THE EFFECT YOUR ACTIONS COULD HAVE ON THE ENVIRONMENT.

18.

JIM: YOU MEAN, LIKE IN A CHRISTMAS CAROL?

19.

SCRIMSHAW JONES (SMILING CRUELLY, MANIACALLY): WITH ALL DUE RESPECT TO MR DICKENS, LET'S CALL THIS ONE A SUMMERTIME SHANTY!

20.

...BEFORE JIM HAS A CHANCE TO DANCE A FAIR IMITATION OF THE HORNPIPE, THE SCENE SHIFTS TO THE SEA FLOOR. HE AND SCRIMSHAW JONES ARE WEARING OLD-FASHIONED DIVING SUITS AND HELMETS WITH AIR HOSES LEADING TO THE SURFACE. THERE IS A TURTLE SWIMMING NEARBY. THE PLASTIC BAG CAN BE SEEN ABOVE, FLOATING, INNOCUOUS AND MENACING...

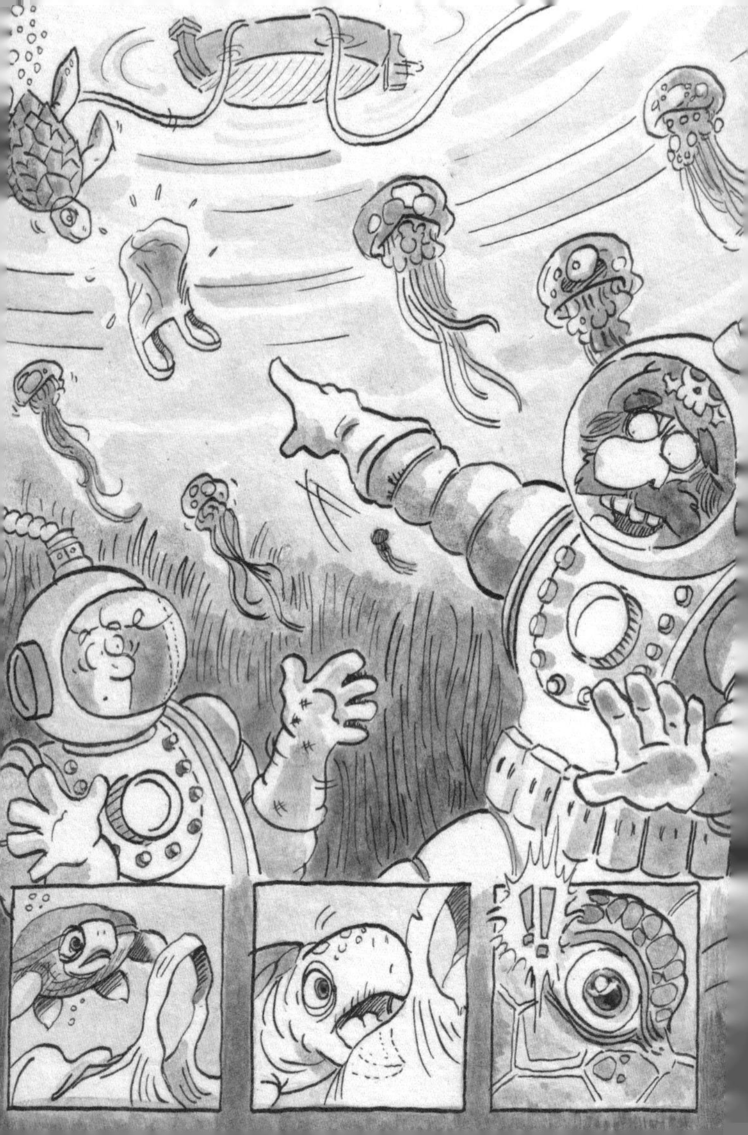

21.
SCRIMSHAW JONES: NOW FROM THAT TURTLE'S PERSPECTIVE, THAT PLASTIC BAG BEARS MORE THAN A PASSING RESEMBLANCE TO A NICE, JUICY JELLYFISH. GUESS WHAT HAPPENS NEXT, KID?

22.
JIM: I DON'T LIKE WHERE THIS VISION QUEST IS GOING...

23.
INDEED, THE SIGHT OF THE DEAD TURTLE GREATLY UPSETS OUR HERO.

24.
SCRIMSHAW JONES: WHAT CAN I SAY, SHIPMATE? THE CAUSE IS OFTEN UNSEEN, BUT THE EFFECT IS USUALLY FELT BY SOMEONE OR SOMETHING. I'M MERELY GIVING YOU THE OPPORTUNITY TO SEE THE CAUSE WITH A VIEW TO YOU REALISING THE ERROR OF YOUR WAYS.

25.
JIM: BUT WHAT CAN I DO?
SCRIMSHAW JONES: KID, IT'S TIME TO THINK GLOBAL AND ACT LOCAL. DO JUST ONE THING. DO IT WELL. ENCOURAGE OTHERS TO DO IT. IN TIME, YOU'LL MAKE AN IMPACT.

26.
JIM: I'M NOT SO SURE...
PIRATE: YOU'LL KNOW WHAT TO DO. OUR TIME'S UP, THIS VISION QUEST IS OVER. DON'T SCREAM TOO LOUD.
JIM: HUH?

27. JIM WAKES, SCREAMING, BUT QUICKLY REALISES THAT HE IS BACK IN HIS OWN BED AND THAT IT WAS ALL A DREAM.

28. BUT OUR HERO HAS LEARNT A LESSON. HE HEARS HIS MOTHER'S VOICE FROM THE NEXT ROOM.

29.
MOM: WHAT'S GOING ON IN THERE? ARE YOU OK?

JIM: I'M GONNA NEED A LIFT TO THE BEACH...

MOM: YOU'RE GOING TO NEED WHAT?

30.
JIM: ... A PAIR OF SAFETY GLOVES...

MOM: HUH?

JIM: ... A CANVAS BAG...

31.
MOM: DON'T MAKE ME SEND YOUR FATHER IN THERE!

JIM: ... AND ADULT SUPERVISION.

MOM: IT'S FIVE O'CLOCK IN THE MORNING. GO BACK TO SLEEP.

32. LATER THAT DAY, JIM HITS THE BEACH WITH A VENGEANCE.

TIME TO THINK GLOBAL AND ACT LOCAL!!
WHAT'S GOTTEN INTO HIM?
NO IDEA...
THE END

Sea Heroes: Jacques-Yves Cousteau

Jacques-Yves Cousteau was a French naval officer who, with Émile Gagnan, invented the modern 'Aqua Lung' diving device in 1943. A keen film maker, Cousteau left the French Navy in 1950 and leased the ship *Calypso* from a descendant of the Guinness family to embark on a series of underwater adventures which are described in his book The Silent World – the film of which won the Palme d'Or trophy at the Cannes Film Festival in 1956. In 1957 he was elected as director of the Oceanographic Museum of Monaco. He created the Cousteau Society for the Protection of Ocean Life in 1975.

Great Shipwrecks in History: Lusitania

The luxury liner *Lusitania* was hit by one torpedo from the German submarine *U-20* during World War One off the Old Head of Kinsale, Co. Cork on the afternoon of the 7th May 1915 and sank within 18 minutes with the loss of 1,191 lives. Many questions have been asked as to why such a large ship could have sunk so quickly. It has even been suggested that she may have been carrying munitions from America to England that were ignited by the German torpedo, creating a second explosion which blew such a big hole in her hull that she could no longer stay afloat.

Great Shipwrecks in History: Titanic

RMS *Titanic* - at the time the largest and most luxurious ship in the world - was built in Ireland at the Harland and Wolf shipyard in Belfast. Her hull, which was divided by vertical walls into 16 separate 'watertight compartments', was designed to be unsinkable. But, unfortunately, these watertight walls did not go all the way to the top of the hull. As luck would have it, her collision with an iceberg during the night of 14th-15th April 1912 opened up five of these compartments to the sea. The weight of water flooding into the hull pulled down the front of the ship so far that water from the forward compartments cascaded backwards into those behind and she sank... with the loss of some 1,500 lives.

Sea Heroes: Robert Ballard

Professor Robert Ballard was the first man to discover the wreck of the famous liner RMS *Titanic*. He did this with the help of the US Navy, who had funded him to search for two lost atomic submarines – *Scorpion* and *Thresher* – on the understanding that any time left over could be used to search for the *Titanic*. Ballard was using the new deep sea underwater robot Argo and discovered *Titanic* – broken in two pieces – on the bottom of the Atlantic in 1985. Ballard went on to explore other famous shipwrecks including the German battleship *Bismarck*, the liner *Lusitania* and President John F. Kennedy's wartime torpedo boat *PT-109*.

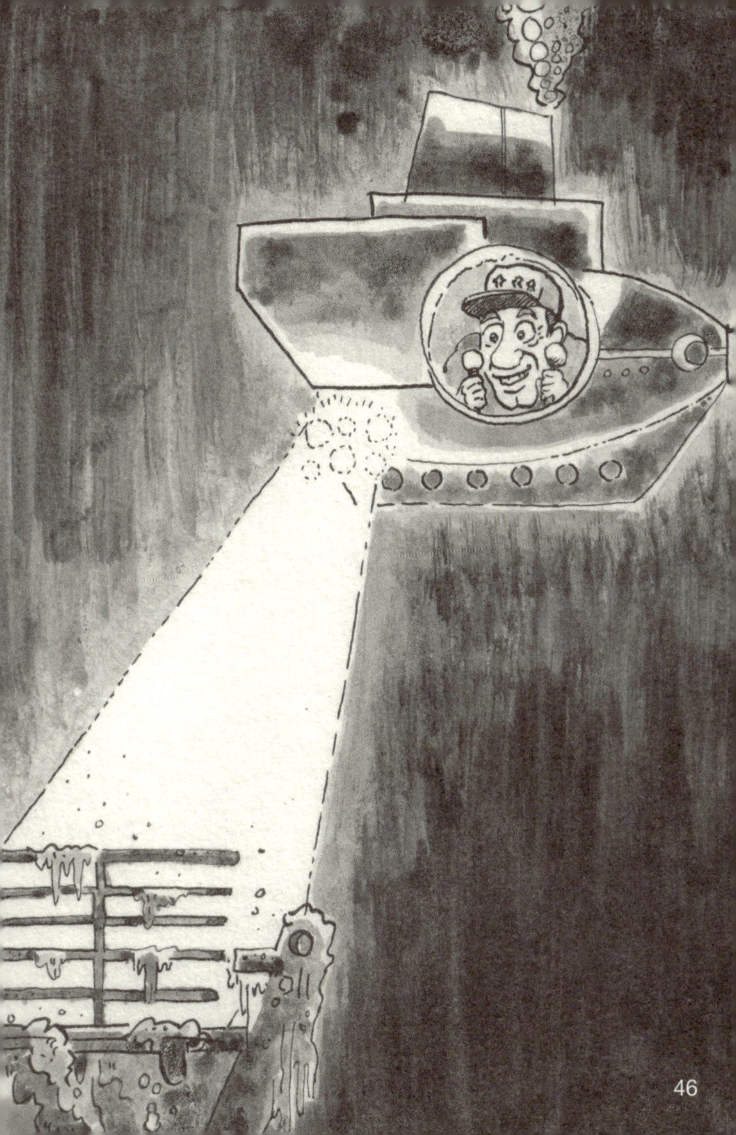

Great Shipwrecks in History: Spanish Armada

In September 1588 a large portion of the 130-strong fleet sent by Philip II to invade England made landfall upon the coast of Ireland.

After being defeated at the naval battle of Gravelines, the Armada attempted to return home through the north Atlantic, but was driven off course toward the west coast of Ireland by violent storms. Up to 24 ships of the Armada were wrecked along the rocky coastline from Antrim to Kerry.

The prospect of a Spanish landing alarmed the Dublin government of Queen Elizabeth I, which prescribed harsh measures for the Spanish invaders and any Irish who might assist them. Many of the survivors of the multiple wrecks were put to death, and the remainder fled across the sea to Scotland. It is estimated that 5,000 members of the fleet perished in Ireland.

10 Songs about the Sea

1. **Yellow Submarine (The Beatles, 1966)**
   A treatise on the insularity of fame or a psychedelic reimagining of the Rhyme of the Ancient Mariner? Either way, one of the catchiest sing-alongs going.

2. **Surfin' U.S.A. (The Beach Boys, 1963)**
   The tune was all Chuck Berry's but the vibe was pure sea and surf bliss courtesy of Brian Wilson.

3. **This is the Sea (The Waterboys, 1985)**
   Emotive and thundering belter from Mike Scott and his crew. 'Once you were tethered/And now you are free/Once you were tethered/Well now you are free/That was the river/This is the sea!'

4. **(My Own Dear) Galway Bay (Frank A. Fahy 1854 1935)**
   More commonly sung in Ireland than Arthur Colahan's similarly-titled composition, which was made famous by Bing Crosby. Dolores Keane produced a renowned version of Fahy's song.

5. **Boney Was a Warrior (Composer Unknown – circa 1825)**
   Stomping sea shanty grudgingly extolling the exploits of one Napoleon Bonaparte.

6. **Rock the Boat (The Hues Corporation, 1974)**
   'We've been sailing with a cargo full of love and devotion' – no Irish wedding is quite complete without this disco smash playing in the early hours.

7. **The SpongeBob Square Pants Them Tune (Mark Harrison, Blaise Smith, Stephen Hillenburg, Derek Drymon, 1999)**
   All together now, 'Who lives in a pineapple under the sea ...'

8. **My Bonnie Lies Over the Sea (Composer Unknown – circa 1750)**
   Perennially popular love song or Jacobite anthem?

9. **The Voyage (Johnny Duhan, 2005)**
   A song about families and love spun through sea and ocean imagery. Quickly becoming a modern classic.

10. **The Wreck of the Edmund Fitzgerald (Gordon Lightfoot, 1976)**
    An epic retelling of the ill-fated SS Edmund Fitzgerald. 'Does anyone know where the love of God goes/When the waves turn minutes to hours?'

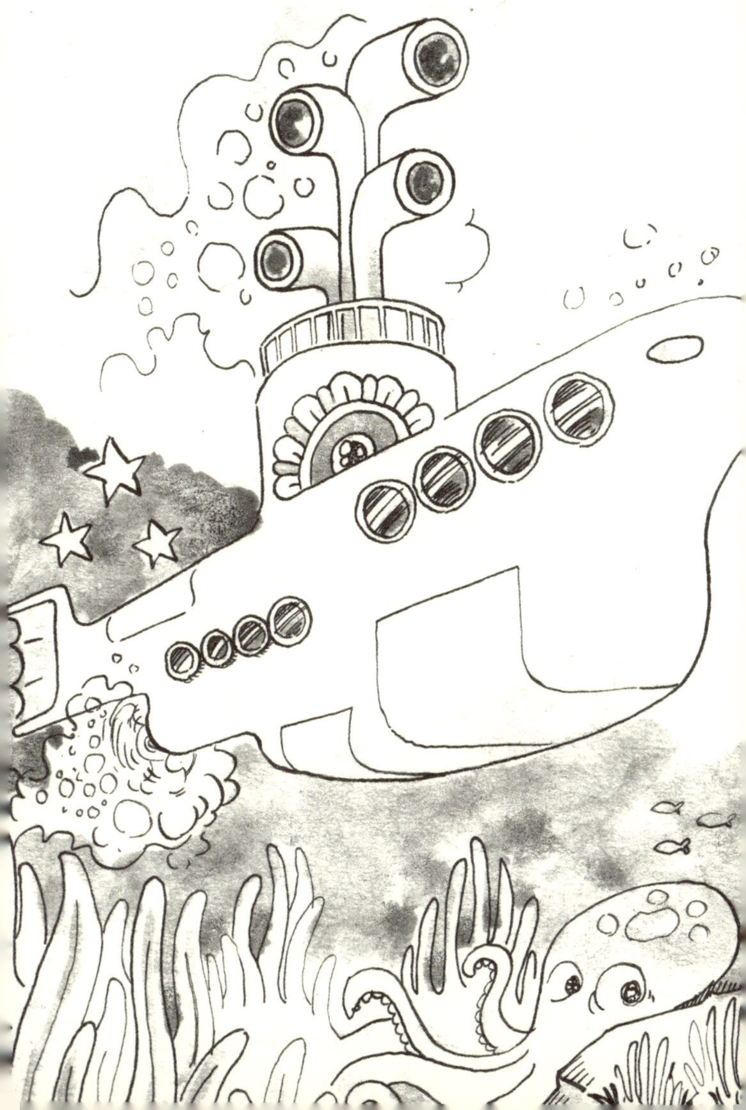

10 Films about the Sea

1. **20,000 Leagues Under the Sea (1954)**
   Old-school undersea sci fi romp, great book too.

2. **The Life Aquatic with Steve Zissou (2004)**
   Bill Murray playfully channels Jacques Cousteau. Jaguar Shark!

3. **Jaws (1975)**
   Boat size doesn't matter.

4. **Mutiny on the Bounty (1962)**
   Many versions, but Brando's is best.

5. **Finding Nemo (2003)**
   Send in the clownfish.

6. **Master and Commander – The Far Side of the World (2003)**
   Imperious adaptation of two of Patrick O'Brien's Aubrey-Maturin novels.

7. **Waterworld (1995)**
   A soggy box office disaster, but is worth revisiting.

8. **Godzilla vs. the Sea Monster (1966)**
   What you see is what you get.

9. **The Hunt for Red October (1990)**
   Connery perfects his Russian accent, underwater.

10. **Cast Away (2000)**
    Hanks upstaged and out-acted by his scene-stealing co-star, the sea.

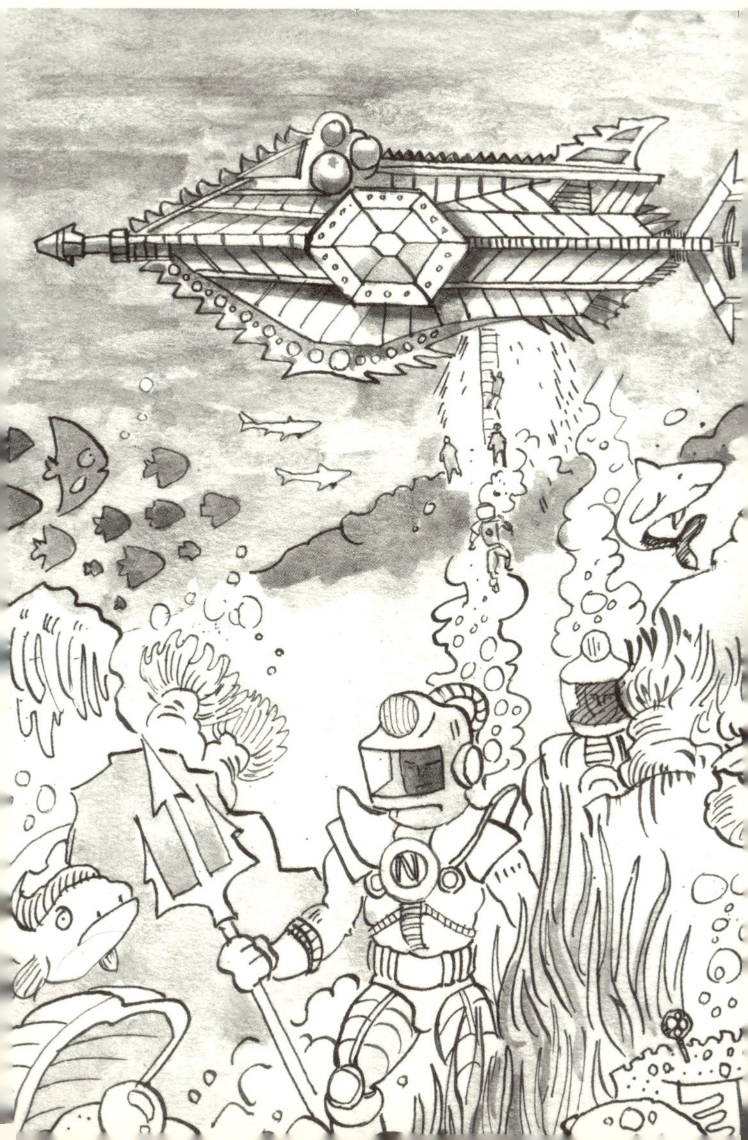

Sea Creatures

Did you know that the following creatures are found in Irish waters?

Sea Turtles

Five marine turtle species appear regularly off the west coast of Ireland but do not come ashore. All are endangered, some critically. These include:
- Green sea turtle
- Loggerhead turtle
- Hawkbill turtle
- Kemp's Ridley
- Leatherback sea turtle

Sharks

Some people are surprised or even nervous to hear there are sharks in Irish waters. In fact, there are 10 species of sharks that use Irish waters.

Some species are migratory and come to Ireland to feed in our rich waters. These include:
- Basking shark
- Blue shark
- Shortfinned Mako
- Six gilled shark
- Porbeagle shark

Other sharks are resident in Ireland year round. These are:
- Spur dogfish
- Greater spotted dogfish

- Lesser spotted dogfish
- Siki dogfish
- Smooth hound

Skates and Rays

Rays are a relative of the shark family. There are many rays in Irish waters, including the thornback ray, blonde ray, sting ray, cuckoo ray, undulate ray, homelyn ray, painted ray, electric ray, common skate and white skate.

Dolphins and Whales (Cetaceans)

Whales and dolphins are known collectively as cetaceans. There are 24 species of cetaceans in Irish waters. Ireland is an important habitat for cetaceans and in 1991, was designated as a sanctuary for whales and dolphins.

Dolphins

Dolphins are found all around the Irish coast. The most familiar species is the bottlenose dolphin. Other species that are found in Ireland are the common dolphin, striped dolphin, risso dolphin, Atlantic white-sided dolphin and whitebeaked dolphin.

Porpoises

The small cousin of the dolphin is the harbour porpoise. They are found in all Irish coastal areas. They feed on crabs, shellfish and fish. Porpoises have flat teeth (spade teeth) to help them crush shells and crab exoskeletons.

Whales

Ireland is one of the best countries in Europe to whale watch. Some of the largest whales, such as the fin whale which measures up to 26m in length, come to Ireland to feed and can be spotted off the coast of Cork (Galley Head), Kerry (Slea Head) and Wexford (Hook Head). Baleen whales don't have teeth, but instead have baleen plates (comb like structures) that they use to sieve water for food. These huge mammals feed on tiny animal plankton (krill) and small fish (herring). Toothed whales have a strong set of teeth to hunt their prey. Toothed whales form groups (pods), unlike the baleen whales, which are usually found alone or in small groups.

Baleen whales found in Irish waters:
- Blue whales
- Fin whales
- Humpback whales
- Sei whales
- Minke whales

Toothed whales found in Irish waters:
- Sperm whale
- Pygmy sperm whale
- Killer whale
- False killer whale
- Sowerby's beaked whale
- Northern bottlenose whale
- Long-finned pilot whale

Seals

There are two native species of true seals in Irish waters, the grey seal and the common (harbour) seal. The grey seal is more abundant than common seal, and is found all over the Irish coastline. The common (harbour) seals are found in more sheltered areas including Carnsore Point (Co Wexford), Clew Bay (Co Mayo) and Tralee Bay (Co Kerry).

Irish Seahorses

Seahorses are commonly found in shallow, tropical and temperate coastal water habitats such as estuaries, coral reefs and coastal sea grasses. Two seahorse species are indigenous to the coast of Ireland: *Hippocampus guttulatus* (the many-branched seahorse), and *Hippocampus hippocampus* (the short-snouted seahorse). Both these species are classified as endangered.

Sea Heroes: Eugenie Clark

Eugenie Clark, sometimes referred to as "The Shark Lady", was an American fish scientist known for her research on poisonous tropical fish and the behaviour of sharks. She was a pioneer in the field of scuba-diving for research purposes.

Clark published the first scientific study demonstrating that sharks can be trained (she taught them to release food by hitting targets with their snouts) and discovered the first effective shark repellent in secretions from a flatfish called Moses sole that lives in the Red Sea. She also ventured into undersea caverns off Mexico's Yucatán Peninsula to find "sleeping sharks" suspended in the water, a discovery that upended scientists' belief that sharks had to keep moving to breathe.

> **"When you see a shark underwater, you should say how lucky I am to see this beautiful animal in its environment."**
>
> Eugenie Clark

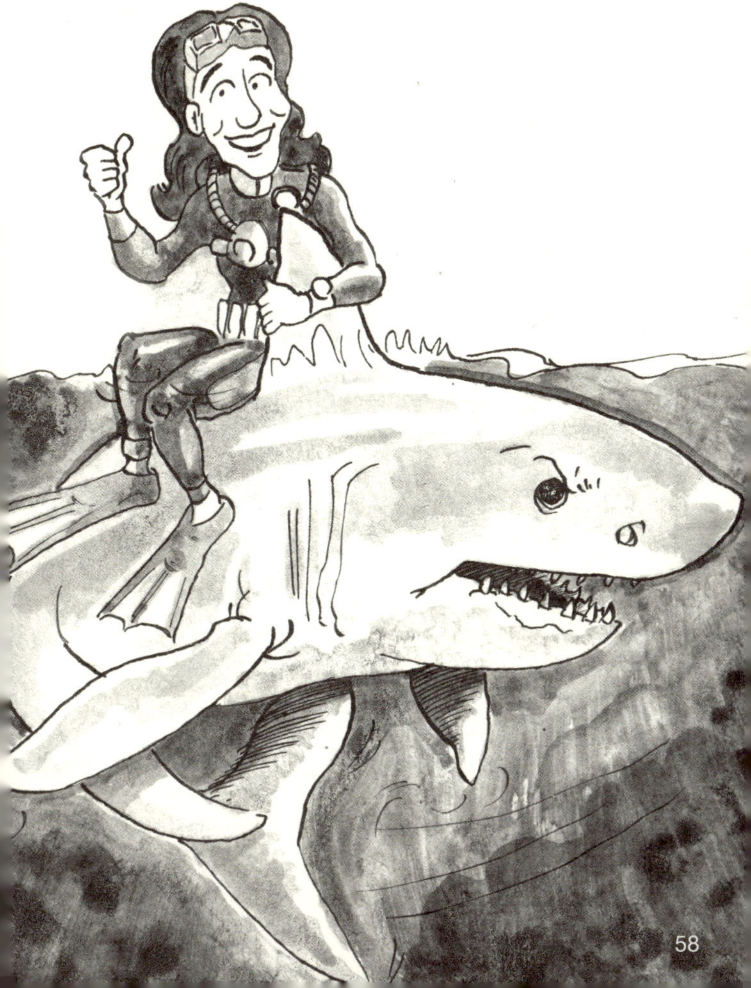

Sea for Society Quiz

Test your sea knowledge! This quiz will be taken by citizens all across Europe as part of the Sea for Society "Collective Action".

Q1: Where does most of the oxygen we breathe come from?
A. Trees in equatorial rain forests
B. Microscopic plants in the sea
C. Grass in fields and meadows

Q2: Where does most of our fresh water come from?
A. Evaporation of freshwater from the sea to create rainclouds, which in turn rain down onto the land
B. The melting of polar ice
C. Floods

Q3: Why is it important to preserve marine biodiversity (i.e. the number of different types of animals and plants living in the sea)?
A. An extensive marine biodiversity allows us a wide range of animals and plants to use in making fertilisers and other products
B. Marine animals and plants are a rich source of raw materials for medicines and other health products
C. Both of the above

Q4: What percentage of the world's edible protein comes from the sea?
A. 3%
B. 10%
C. 40%

Q5: Which is the most visited type of holiday destination in the world?
A. Mountains
B. Countryside
C. Seaside

Q6: Which ancient monarch built of the first tools to explore the marine environment?
A. King Arthur
B. Henry VIII
C. Alexander the Great

Q7: Where was the first underwater telegraphic cable laid?
A. In the Bosphorus Strait in 1915
B. Between France and England in 1850
C. Between Sweden and Denmark in 1947

Answers can be found on page 65

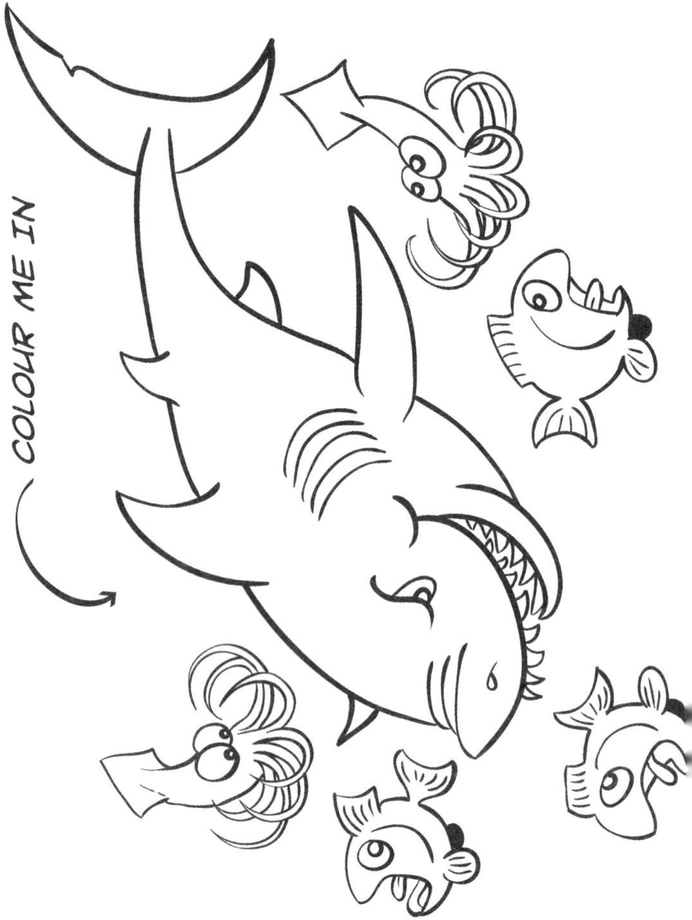
COLOUR ME IN

COLOUR ME IN

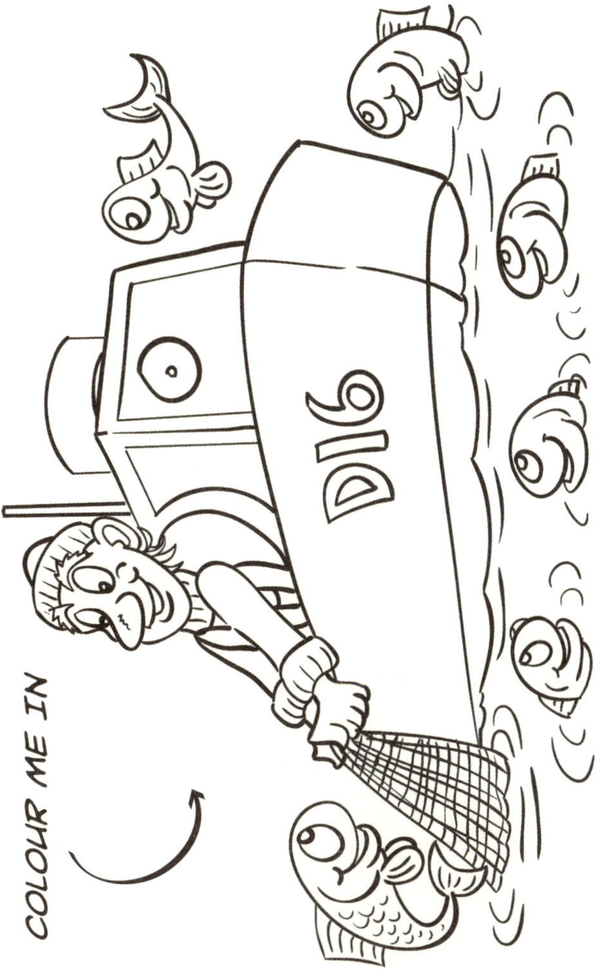

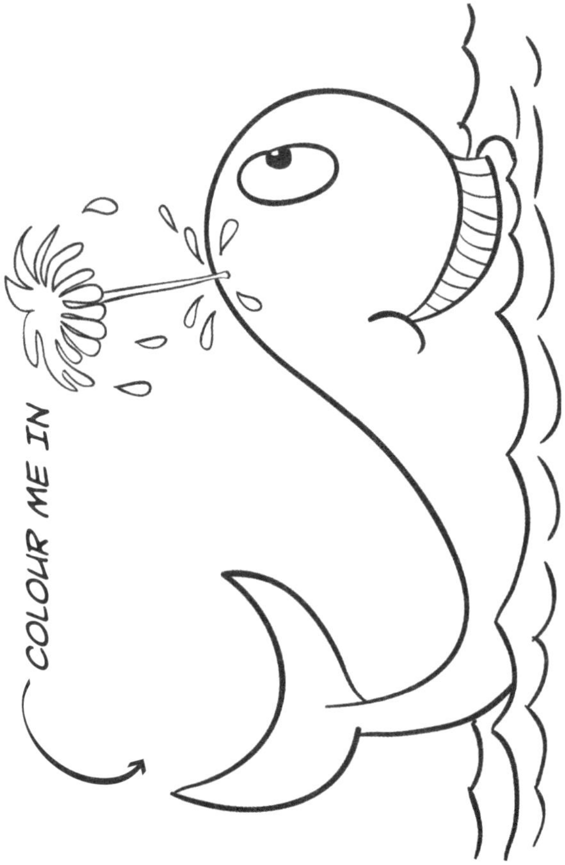
COLOUR ME IN

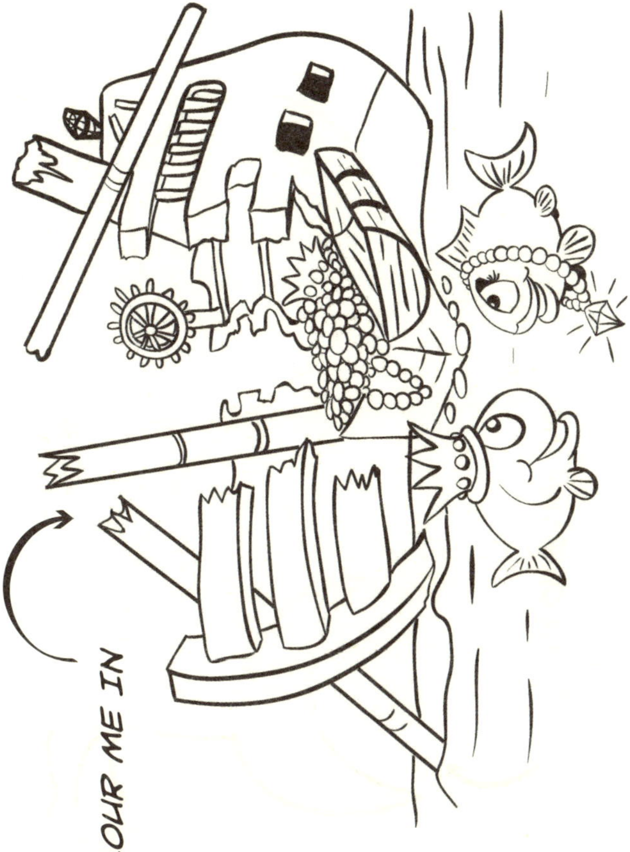

COLOUR ME IN

Quiz Answers

Answer 1: B. Microscopic plants in the sea. (Marine phytoplankton). These produce approximately half of all the oxygen in our atmosphere.

Answer 2: A. Evaporation of freshwater from the sea. Most of our freshwater comes from the sea by way of rainclouds. Sunlight evaporates an estimated 16 billion litres of freshwater every second from the world's oceans – equal to an incredible 502,800 cubic kilometres each year. This then falls back to the surface of our planet as rain – both on the land and the sea.

Answer 3: C. Both of the above. In addition – the sea is also a rich source of minerals, as well as fossil fuels and renewable energy from tides, winds and waves. As technology develops into the future, we may discover other ocean resources which will learn how to harvest these resources in new and sustainable ways.

Answer 4: B. 10%. While national diets vary in the amount of marine protein eaten, when averaged out across the world, 10% of edible protein eaten is derived from the sea.

Answer 5: C. Seaside. Marine leisure and tourism is the backbone of many local economies in Europe's coastal regions and across the world.

Answer 6: C. Alexander the Great. The first diving bell was built by Alexander the Great on this way back from the Indies in 325 BC to view marine creatures. This primitive underwater vehicle – which he named 'Colympha' - was the ancestor of modern diving bells.

Answer 7: **B. Between France and England in 1850.** In 1850, a collaboration between England and France lead to the installation of the first underwater telegraph cable – linking southern England and the Cap Gris-Nez in France.

Do Just One Thing Checklist

✓	ACTION
☐	1. Dispose of plastic products responsibly – do not throw them in the sea or leave them on the beach
☐	2. Volunteer to take part in a local beach clean
☐	3. Only eat fish that has been caught from 'Sustainable Fisheries'
☐	4. Always read the labels on pet food containers to make sure that it contains only sustainably caught marine proteins
☐	5. Leave your car at home
☐	6. Turn down your heating
☐	7. Have a shower instead of a bath
☐	8. Don't buy holiday souvenirs that exploit the oceans such as dried seahorses, seashells and corals
☐	9. Learn more about the sea
☐	10. Spread the word